Danilo Cardoso Ferreira

# Spectroscopic processes with lignins

Danilo Cardoso Ferreira

# Spectroscopic processes with lignins

## UV-vis, FTIR and Impedance Measurements

ScienciaScripts

**Imprint**



Cover image: www.ingimage.com

This book is a translation from the original published under ISBN 978-613-9-65513-7.

Publisher:
Sciencia Scripts
is a trademark of
Dodo Books Indian Ocean Ltd. and OmniScriptum S.R.L publishing group

120 High Road, East Finchley, London, N2 9ED, United Kingdom
Str. Armeneasca 28/1, office 1, Chisinau MD-2012, Republic of Moldova, Europe
Printed at: see last page
**ISBN: 978-620-7-78034-1**

# Summary

I dedicate this work to my wife *Michelle Prado*.

*"Faith needs the light of science so that it doesn't become blind (fideism) and become fanatical, fundamentalist and dangerous, as we have seen today; science, on the other hand, needs faith so that it doesn't fall into rationalism and put its discoveries at the service of evil."*

*(Science and Faith in harmony, p. 26)*

Part I

# PVD Films of Methanol Lignin and UV-vis Absorption Spectroscopy Measurements

## Chapter 1

# Introduction and justification (with a summary of the literature)

In this scientific initiation project, ultra-thin films (nanometers thick) of lignins extracted from sugar cane bagasse will be manufactured using the physical vapor deposition (PVD) technique, with the aim of advancing the study of the molecular properties of lignins. The main information obtained in this project refers to the possibility of manufacturing quality PVD films in terms of the chemical integrity of the molecule, since they will be heated, as well as optimizing the control of the thickness, homogeneity and molecular arrangement of these lignins in the PVD films. Film growth was monitored using ultraviolet and visible (UV-vis) absorption spectroscopy, while chemical integrity was checked using Fourier transform infrared (FTIR) absorption spectroscopy. The lignin studied was extracted from sugar cane bagasse via a "modified organosolve" process using different organic solvents. The modification to the extraction process consists of subjecting the solvents to supercritical fluid conditions during extraction. This IC project is being carried out at the DFQB of the FCT/UNESP in Presidente Prudente and will have the collaboration of post-doctoral student D. Pasquini, who is responsible for the extraction of the lignins and will be supervised by Professor Dr. A. Apri- gio S. Curvelo of the IQSC/USP, our scientific collaborator.

This report describes the activities carried out between January and July 2008 working with lignin extracted from sugar cane using the solvent methanol. The following activities were planned for this period: i) to make PVD films of butanol lignin with different thicknesses, monitoring this growth via UV-vis absorption; ii) to investigate the chemical integrity of the lignins in the PVD films via FTIR absorption. Objectives (i) and (ii) were fully met, but using lignin extracted from sugar cane bagasse with the solvent methanol. The change from butanol lignin to methanol was made for two reasons: we had little butanol lignin material and it was the only one that made it possible to produce LB films. In addition, we carried out optical microscopy measurements of the butanol lignin PVD films to check their morphology on a microscopic scale and tested them in the UV-vis washing process to investigate the possible loss of material, since these films should be applied in sensory units for detecting analytes in liquid media. This report is organized as follows. Section II provides the theoretical concepts, Sections III and VI describe the experimental procedure and the results and discussion. Section V, in particular, contains the conclusive information obtained on lignins via PVD films.

## Chapter 2

# Theoretical concepts

## *2.1* PVD films

PVD films are made by evaporating the powder in a metal, cylindrical crucible, which has a small hole at the top through which the evaporated material escapes. The crucible is heated by the passage of an electric current, the intensity of which is generally no more than 2.0 A (10 V), which implies temperatures of less than 200° C for Ta (Tantalum) crucibles. The evaporation process is facilitated by the vacuum (≈ $10^{-6}$ Torr) inside the hood where the whole process takes place. The substrate is positioned above the crucible containing the material, as is the thickness controller (quartz crystal scale). A shutter positioned between the metal crucible and the substrate is used to protect the substrate while the evaporation rate is regulated. Once the deposition rate has been controlled, the thickness controller is "reset", the shutter is opened and the PVD film manufacturing process begins until the desired thickness is reached. At this point the shutter is closed and the current is interrupted [1-2].

## *2.2* UV-vis and FTIR absorption spectroscopy

UV-vis and FTIR absorption spectroscopy are techniques that make it possible to characterize the material by the incidence of electromagnetic radiation at different frequencies, which can take the molecules into an excited state of energy by absorbing the radiation. Electronic transitions involve the absorption of electromagnetic radiation whose energy is in the visible or ultraviolet region, UV-vis spectroscopy. The frequency of the absorbed radiation is a measure of the energy required for the electronic transition, while the intensity depends on the probability of this transition occurring [3-4]. In film research, the UV-vis technique is widely used to verify the growth of films, since, for a deposition with controlled thickness, the absorbance must increase linearly with the increase in the number of layers deposited, following Beer's law [5]. Determining the group responsible for absorbing UV-vis radiation and the inter-wavelength range at which absorption occurs are also important pieces of information [6].

In FTIR spectroscopy, the incident radiation is absorbed when the energy of this radiation corresponds to the energy difference between two vibrational levels of the molecule. FTIR spectroscopy makes it possible to investigate intra- and inter-molecular interactions through the absorption bands of the spectrum relating to vibrational transitions, as well as making it possible to identify the functional groups of the substances to be detected. Another potential use of FTIR is to use surface selection rules [2] to obtain information about the structural arrangement of molecules in a film due to the possible anisotropy induced during its manufacture or in subsequent treatments [7]. This study involves comparing the FTIR spectra of the material obtained using KBr pellets, film on a substrate transparent to IR radiation and film on a surface reflecting IR radiation.

## Chapter 3

# Experimental procedure

The lignin extraction process used in this work was described in the research project of former scholarship holder Gislaine Fellipe Martins. Here we summarize the data relating to the lignin studied, which was extracted using methanol (ML) and has the following molar mass in g/mol: ML (Mn=1150, Mw=1818, Mw/Mn=1.58). These data were determined by GPC using an IR detector.

## 3.1 PVD film production

The PVD films of ML lignin were manufactured using a Boc Edwards model 306 vacuum evaporator. Due to the limited amount of material available, it was decided to control each film deposition on the substrate using mass and thickness ratios. Approximately 5.0 mg were used in each deposition process, measured on an analytical balance, taking care to note the value of the thickness obtained in each deposition, which is read by a quartz crystal balance attached to the evaporator itself. Table I contains the masses weighed for ML lignin. The intensity of the electric current used to heat the Ta crucible was around 1.95 A. This value was obtained from a thermocouple. It should be noted that only 10 depositions were made for this ML lignin due to the limited material available.

| Deposit | Mass (mg) | Thickness (nm) |
|---|---|---|
| 1ª | 5,4 | 20,9 |
| 2ª | 5,6 | 17,9 |
| 3ª | 5,2 | 24,6 |
| 4ª | 5,4 | 17,7 |
| 5ª | 5,6 | 18,3 |
| 6ª | 5,2 | 21,2 |
| 7ª | 5,7 | 19,5 |
| 8ª | 5,4 | 17,7 |

| 9[a] | 5,5 | 25,7 |
|---|---|---|
| 10[a] | 5,6 | 26,1 |

Table 3.1: Lignin Methanol

# 3.2 Casting film production

The casting films were prepared by spreading 400 $\mu L$ of lignin solution in tetrahydrofuran (THF) at a concentration of 0.8 mg/mL onto a zinc sealing (ZnSe) substrate. After total evaporation of the solvent under a controlled atmosphere (hood with exhaust air flow), the film formed is known as the casting film.

# 3.3 UV-vis and FTIR

The growth of the ML lignin PVD films was monitored by UV-vis absorption using a Varian Cary 50 spectrophotometer, scanning from 190 to 800 nm and using a quartz slide with no film deposited as a reference (baseline). The study of film loss to water after successive washes of the film was also monitored using UV-vis spectroscopy. The FTIR measurements of the PVD films deposited on ZnSe substrates were carried out using a Bruker Vector 22 spectrometer, with a spectral resolution of 4 cm-1 and 128 scans in transmission mode. Optical microscopy measurements were carried out using a Leica microscope with a 50X objective.

# Chapter 4

# Results and discussion

## 4.1 UV-vis absorption spectroscopy

UV-vis spectroscopy was used to check the growth of the PVD films. Figure 1 shows the UV-vis absorption spectra obtained for consecutive depositions of ML lignin, the masses of which are given in Table I. The growth of the PVD film was monitored by the maximum absorption band at around 280 nm, which is attributed to the $\pi \dashrightarrow \pi^*$ transition of the phenyl group [5, 8]. It is worth noting that the UV-vis spectra obtained for the PVD films are very similar to those obtained for Langmuir-Blodgett (LB) films [9], suggesting no thermal degradation of the lignins, at least in relation to the phenyl group.

The inset in Figure 1 shows that there is a linearity between the maximum absorbance value at 280 nm and the film thickness expressed in nm (or amount of material deposited in mg). This linearity indicates that with each evaporation process a similar amount of material was deposited on the substrate. A reduction of approximately 18 to 20 nm was found for every 5.0 mg of lignin.

The linearity observed for the absorbance as a function of the mass deposited or

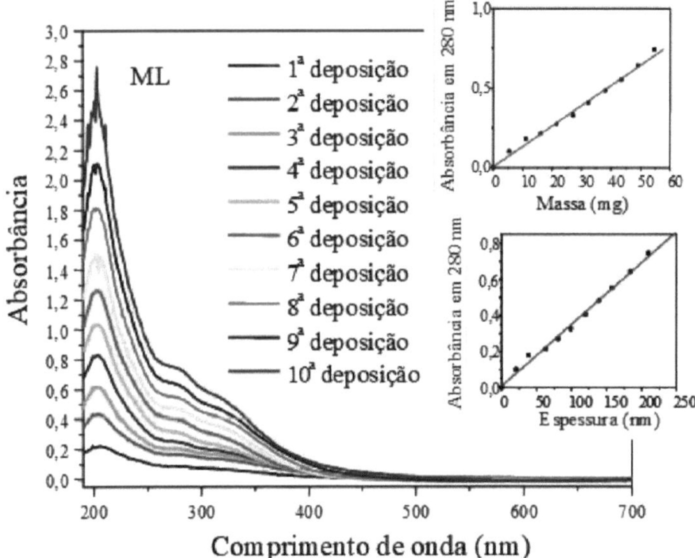


Figura 4.1: UV-vis absorption spectra of the ML lignin PVD film made from successive depositions on quartz sheets. Upper inset: absorbance at 280 nm vs mass in mg of deposited material. Lower inset: absorbance at 280 nm vs thickness in nm.

The thickness of the PVD films (insets in Figure 1) indicates that fairly close quantities of material are evaporated with each deposition process, but this does not guarantee that the material is deposited homogeneously on the substrate. Therefore, optical microscopy measurements were carried out to make inferences about the morphology of these PVD films on a microscopic scale. An image is shown in Figure 2 for the ML lignin PVD films made with 10 depositions on quartz substrates. It can be seen that on a microscopic scale the films obtained are quite homogeneous, which is an important factor in terms of the reproducibility of the results when applying these films as sensors via electronic language, for example. With a view to the future application of these films in sensory units, it was checked whether there was any loss of material during the washing process, which is also an important factor in guaranteeing the reproducibility of the results when applying these films. The ML lignin PVD films with maximum depositions on a quartz substrate were subsequently washed in water.

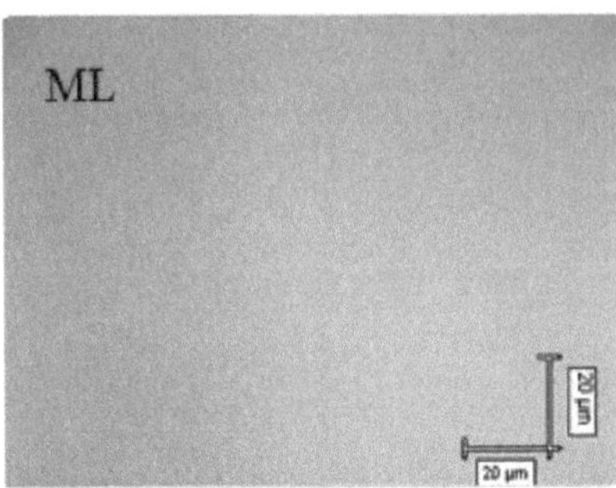


Figura 4.2: optical microscopy images of ML lignin PVD films deposited on quartz foil.

following the procedure adopted when working with electronic tongue sensors. The washing process consisted of immersing the film in ultrapure water contained in a beaker and on a magnetic stirrer. Setting the stirrer in motion created a gentle vortex to which the substrate was subjected for 5 minutes. Before and after each wash, a UV-vis absorption spectrum of the film was obtained, which is shown in Figure 5. It is possible to notice a significant loss of material during the wash, which, in principle, would make it impossible to apply these PVD films as sensory units if you wanted to reuse the sensors after a certain measurement. However, this loss most likely occurs because the film is relatively thick (many depositions -→ 150 to 250 nm), leading to a loss of material closer to the surface, i.e. further away from the substrate. There is a tendency for the loss of material to stabilize towards the "layers" closest to the substrate (lower absorbance). Considering that in the case of the sensory units the PVD films are deposited at a thickness of around 10 nm, the expectation is that the loss will be very small, as has already been observed for LB films [9]. In any case, these measurements need to be repeated under the conditions in which the sensory units are manufactured (PVD films 10 nm thick).

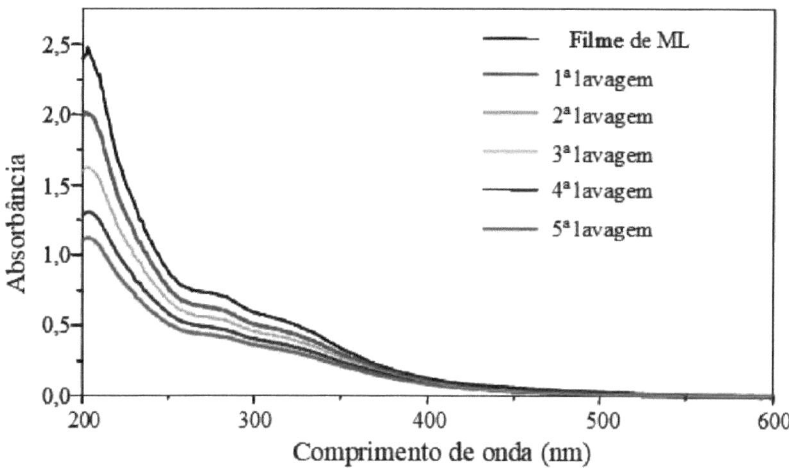


Figura 4.3: UV-vis absorption spectra, before and after successive washes, of ML lignin PVD films deposited on quartz.

## 4.2 Infrared absorption spectroscopy

Infrared absorption spectroscopy measurements were carried out in order to investigate the thermal stability of the lignin molecules subjected to the vacuum evaporation process to form the PVD films. Figure 4 shows the FTIR spectrum obtained for the PVD film and ML lignin casting. The assignments of the main absorption bands are given in Table II [9]. The agreement in terms of wavenumber for both PVD and casting films shows that the PVD film manufacturing process does not induce changes in the molecular structure of the lignin. Furthermore, the main bands observed for the PVD films are the same as those reported in the literature for lignins in general [1012], and for these lignins extracted from sugar cane via the supercritical process [13]. The difference in the relative intensity of some bands when comparing the FTIR spectra for PVD and casting films may be related to the different organization that lignins can acquire in thin films depending on the manufacturing process, as discussed below.

Figure 5 shows the FTIR spectra of the ML lignin PVD film and the butanol lignin (BL) LB film. Thus, two considerations can be made: i) the ML molecules are not randomly structured, as is the case with cast films,

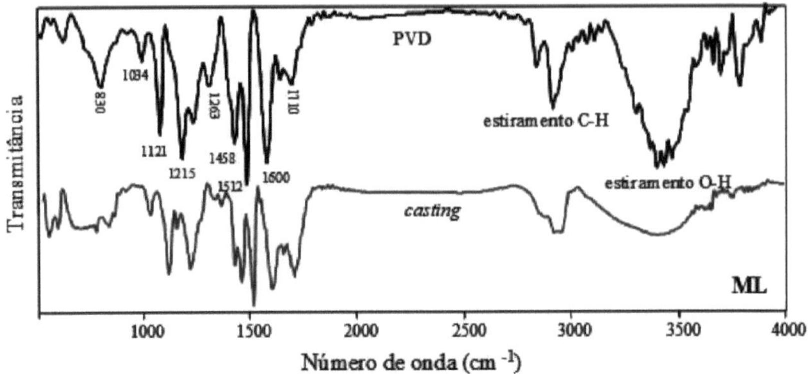


Figura 4.4: FTIR spectra of PVD films and ML lignin casting, both de-positioned on ZnSe.

| Center (cm 1) | Duties |
|---|---|
| 830 | Aromatic C-H (out-of-plane deformation) |
| 1034 | Aromatic C-H (in-plane deformation) and alcoholic C-O (deformation) |
| 1121 | C-O-C ether (deformation) |
| 1215 | C-O-C ether (deformation) |
| 1263 | Guaiacyl ring (deformation) |
| 1458 | C-H (deformation) |
| 1512 | Aromatic ring (C=C stretch) |
| 1600 | Aromatic ring (C=C stretch) |
| 1710 | C=O carbonyl (stretching) |

Table 4.1: Wave numbers and assignments for the main FTIR bands

as shown by the difference in the relative intensity of the bands in the spectra of the PVD and casting films; ii) the organization of the ML in the PVD films is different from that of the BL lignin in the LB films [9], which is also indicated by the difference in the relative intensity of the bands in the spectra for both the PVD and LB films. In the case of LB films, the BL lignins are preferentially positioned with the benzene rings parallel to the substrate surface. In the case of ML PVD films, it will be necessary to deposit them on Au or Ag substrates (mirrors) in order to determine their organization. This procedure is carried out by comparing the FTIR spectra in the transmission and reflection modes, since in these cases the electric field of the incident radiation is polarized parallel and perpendicular to the plane of the substrate, respectively. Knowing this and considering that the absorption intensity is given by the scalar product, where E is the electric field of the incident radiation *I* = ~ -~' the variation produced in the dipole moment of the molecule by the incident radiation, the organization of the molecule in the film can be determined. This is a procedure that has been carried out in our research group for different materials, in addition to lignins [9, 14], such as phthalocyanines [1], perylene [2, 15] and polymers [16-17].

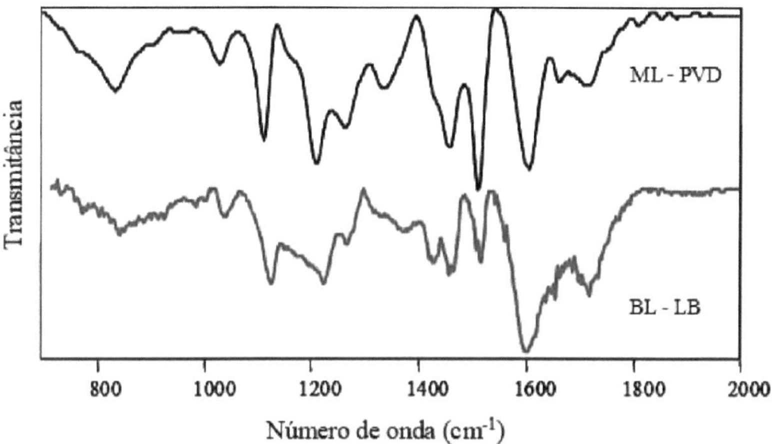


Figura 4.5: FTIR spectra of the ML lignin PVD film and the BL lignin LB film, all deposited on ZnSe.

## Chapter 5

# Experimental procedure

PVD film with controlled thickness on a nanometric scale was obtained using lignin extracted with the solvent methanol (ML), which was verified through UV-vis measurements for the growth of the film with different numbers of depositions. In morphological terms, these films are homogeneous on a microscopic scale and AFM measurements will be carried out to check the morphology on a nanometric scale. Using FTIR absorption spectroscopy, it was possible to investigate the molecular integrity of the lignin in the PVD films and verify that it does not degrade at the temperature used to manufacture them. In addition, via FTIR, it was possible to observe that ML lignin has a different molecular organization in the PVD films and that this organization is different from that found in the LB film. The determination of this molecular organization will be investigated via FTIR measurements combining transmission and reflection modes. Finally, the results obtained in this project show that the PVD film of this lignin could become a sensory unit for liquid systems known as electronic tongues.

**Part I References**

1. L. Gaffo, C.J.L. Constantino, W.C. Moreira, R.F. Aroca, O.N. Oliveira Jr., Journal of Raman Spectroscopy, 33 (2002) 833.

2. P.A. Antunes, C.J.L. Constantino, J. Duff, R. Aroca, Applied Spectroscopy, 55 (2001) 1341.

3. R.M. Silverstein, G.C. Bassler, T.C. Morril, Spectrometric identification of organic compounds, Guanabara Koogan (1994).

4. D. Campbell, J.R. White, Polymer characterization - physical techniques, Chapman & Hall (1989).

5. J.B. Lambert, H.F. Shurvell, D.A. Lightner, R.G. Cooks, Organic structural

spectroscopy, Prentice-Hall (1998).

6. P.A. Antunes, C.J.L. Constantino, R.F. Aroca, Langmuir, 17 (2001) 2958.

7. A.M. Bradshaw, E. Schweizer in Spectroscopy of Surface, R.J.H. Clark, R.E. Hester (Eds.), John Wiley & Sons (1988).

8. S.Y. Lin, C.W. Dence, Methods in Lignin Chemistry, Springer-Verlag, Berlin (1992).

9. A.A. Pereira, G.F. Martins, P.A. Antunes, R. Conrrado, D. Pasquini, A.E. Job, A.A.S. Curvelo, M. Ferreira, A. Riul Jr., C.J.L. Constantino, Langmuir, 23 (2007) 6652.

10. S.A. Gundersen; M.H. Ese; J Sjoblom, Journal of Colloids and Surfaces A, 182, (2001) 199.

11. F. Martin-Dupont; V. Gloaguen; M. Guilloton; R. Granet; P. Krausz, Journal of Environmental Science and Health Part A, 41, (2006) 149.

12. L. Dupont; J. Bouanda; J. Dumonceau; M. Aplincourt, Journal of Colloid and Interface Science, 263, (2003) 35.

13. G.F. Martins, A.A. Pereira, B.A. Straccalano, P.A. Antunes, D. Pasquini, A.A.S. Curvelo, M. Ferreira, A. Riul Jr., C.J.L. Constantino, Sensors and Actuators B, 129, (2008) 525.

14. D. Pasquini, D.T. Balogh, P.A. Antunes, C.J.L. Constantino, A.A.S Curvelo, R.F. Aroca, O.N. Oliveira Jr., Langmuir, 18 (2002) 6593.

15. D. Volpati, A.E. Job, R.F. Aroca, C.J.L. Constantino, The Journal of Physical Chemistry B, 112 (2008) 3894.

16. M. Ferreira, C.J.L. Constantino, C.A. Olivati, M.L. Vega, D.T. Balogh, R.F. Aroca, R.M. Faria, O.N. Oliveira Jr, Langmuir, 19 (2003) 8835.

17. K. Wohnrath, C.J.L. Constantino, P.A. Antunes, P.M. dos Santos, A.A. Batista, R.F. Aroca, O.N. Oliveira Jr, The Journal of Physical Chemistry B, 109 (2005) 4959.

## Part II

# PVD Films of Lignin Methanol, Ethanol, Propanol and Butanol with UV-vis Absorption Spectroscopy Measurements

# Chapter 6

# Introduction and justification

In this scientific initiation project, ultra-thin films (nanometers thick) of lignins extracted from sugar cane bagasse will be manufactured using the physical vapor deposition (PVD) technique with the aim of advancing the study of the molecular properties of lignins. The main information obtained in this project refers to the possibility of manufacturing quality PVD films in terms of the chemical integrity of the molecule, since they will be heated, as well as optimizing the control of the thickness, homogeneity and molecular arrangement of these lignins in the PVD films. Film growth was monitored by ultraviolet and visible (UV-vis) absorption spectroscopy, while chemical integrity was checked by Fourier transform infrared (FTIR) absorption spectroscopy. The lignins studied were extracted from sugar cane bagasse via the "modified organic solvent" process using different organic solvents. The modification to the extraction process consists of subjecting the solvents to supercritical fluid conditions during extraction. This IC project is being carried out at the DFQB of the FCT/UNESP in Presidente Prudente and will have the collaboration of post-doctoral student D. Pasquini, who is responsible for the extraction of the lignins and supervised by Prof. D. Pasquini.

Dr. A. Aprigio S. Curvelo from IQSC/USP, our scientific collaborator.

## 6.1 Justification

This report describes the activities carried out between August 2008 and September 2009 working with lignins extracted from sugar cane bagasse using the solvents methanol, ethanol, propanol and butanol. The following activities were planned for this period: i) fabricate PVD films of the four lignins with different thicknesses, monitoring this growth via UV-vis absorption spectroscopy; ii) investigate the chemical integrity of the lignins in the PVD films via FTIR absorption spectroscopy;

iii) check the adhesion of these PVD films on glass substrates, via UV-vis absorption, when subjected to washing processes similar to those carried out with the sensory units; iv) determine the organization of the lignin molecules in the PVD films via FTIR absorption in the transmission and reflection modes; v) through objectives (i)-(iv), generate contributions to the understanding of the molecular arrangement of lignins in PVD films and optimize the experimental parameters for the manufacture of PVD films of these lignins. Objectives (i) to (v) were fully met, but the FTIR reflection mode measurements were not carried out, as the FTIR transmission mode spectra were compared with the spectrum of the lignin powder, which did not jeopardize the objective of this work. In addition, optical microscopy measurements were taken of the PVD films of the four lignins to check their morphology on a microscopic scale. This report is organized as follows. Section II provides theoretical concepts on PVD films and UV-vis and FTIR absorption spectroscopy. Sections III and VI describe the experimental procedure and the results and discussion, respectively. Section V contains the conclusions of the work on lignins in PVD films.

The study of lignin is interesting due to its richness in aromatic systems (benzene rings), since it can represent an alternative source not only of benzene, currently obtained from petroleum, but also an important raw material in the polymer industry and other derivatives used as inputs in chemical industries. Research has been carried out with the aim of enabling lignins to be used in more technologically advanced applications, based on a better understanding of the properties of this complex and abundant raw material. Research using computer simulations [1] is one alternative. Investigation via ultra-thin films has been another possibility and some articles can be found in the literature [2], including Langmuir films [3-7], LB [7-12], self-assembled (layer by layer) [13], spin-coating [14, 15] and dip-coating using specific substrates [16]. These techniques have made it possible to obtain information on the molecular level of these ma terials, mainly regarding their morphology, the thickness of the monolayer and the area of the molecules that make it up, as well as possible applications. This information can contribute to a better understanding of delignification mechanisms by analyzing solubilized lignin under different experimental conditions [3, 5, 8, 17]. The importance of manufacturing thin films lies in the fact that a large proportion of technological devices based on organic materials are made using these materials in the form of thin films. In addition, the molecular architecture of these thin films plays an important role in their electrical and optical properties and, consequently, in the performance of the device. In general, the molecular architecture of thin films is understood to be their thickness, molecular organization, morphology and crystallinity.

# Chapter 7

# Theoretical concepts

## 7.1 PVD films

The physical vapor deposition (PVD) process refers to a set of deposition methods in which the material is deposited atom by atom or molecule by molecule. The materials are vaporized from a solid or liquid source in the form of atoms or molecules, and are transported in the form of steam through a low-pressure, gaseous or vacuum atmosphere, solidifying on contact with the substrate [18]. PVD methods normally allow high deposition rates without causing damage to the substrate surface, due to the low energy of the incident species. PVD processes usually have a thickness monitoring system that acts during the process (piezoelectric crystal) [19]. Typically, PVD processes are used to deposit films with thicknesses of a few nanometers to micrometers, and can also be applied to produce many layers. The PVD process can be carried out by vacuum evaporation, sputtering, arc deposition and others.

Vacuum evaporation, used in this work, is a type of PVD process in which the material vaporizes from a thermal source and reaches the substrate with little or no collision with gas molecules in the space between the source and the substrate. Vacuum evaporation can be carried out by resistive heating in which the source material is placed in a metal crucible or in a tungsten filament. The support is then heated by the Joule effect, melting the source material. Although very simple, evaporation by resistive heating has several drawbacks, such as: refractory metals and many polymers cannot be evaporated due to their high melting point; evaporation of the filament material can contaminate the film and the film thickness and alloy composition cannot be precisely controlled [19].

## 7.2 UV-vis and FTIR absorption spectroscopy

UV-vis and FTIR absorption spectroscopy are techniques that make it possible to characterize the material by the incidence of electromagnetic radiation at different frequencies, which can take the molecules into an excited state of energy by absorbing the radiation. Electronic transitions involve the absorption of electromagnetic radiation whose energy is in the visible or ultraviolet region, UV-vis spectroscopy. The frequency of the absorbed radiation is a measure of the energy required for the electronic transition, while the intensity depends on the probability of this transition occurring [20-22]. In film research, the UV-vis technique is widely used to verify the growth of films, since, for a deposition with controlled thickness, the absorbance should increase linearly with the increase in the number of layers deposited, following Beer's law [22]. Determining the group responsible for absorbing UV-vis radiation and the inter-wavelength range at which absorption occurs are also important pieces of information [23].

In FTIR spectroscopy, the incident radiation is absorbed when the energy of this radiation corresponds to the energy difference between two vibrational levels of the molecule. FTIR spectroscopy makes it possible to investigate intra- and inter-molecular interactions through the absorption bands of the spectrum relating to vibrational transitions, as well as making it possible to identify the functional groups of the substances to be detected. Another potential use of FTIR is to use surface selection rules [24] to obtain information about the structural arrangement of molecules in a film due to the possible anisotropy induced during its manufacture or in subsequent treatments [25]. This study involves comparing the FTIR spectra of the material obtained using KBr pellets, film on a substrate transparent to IR radiation and film on a surface reflecting IR radiation.

# Chapter 8

# Experimental procedure

The lignin extraction process used in this work was described in the research project. Here we summarize the data on the lignins studied, which were extracted using methanol, ethanol, propanol and butanol and have the following molar masses in g/mol: methanol ($Mn$ = 1150, $Mw$ = 1818, $Mw/Mn$ = 1.58), ethanol ($Mn$ = 1185, $Mw$ = 1971, $Mw/Mn$ = 1. 66), propanol ($Mn$ = 1082, $Mw$ = 1647, $Mw/Mn$ = 1. 52) and butanol ($Mn$ = 1062, $Mw$ = 1594, $Mw/Mn$ = 1. 50).

## 8.1 PVD film production

PVD films are made by evaporating powder in a metal, cylindrical crucible, which has a small hole at the top through which the evaporated material escapes. The crucible is heated by the passage of an electric current, the intensity of which is generally no more than 2.0 A (10 V), which implies temperatures of less than 200° C for Ta crucibles. The evaporation process is facilitated by the vacuum (≈ $10^{-}$ 6 Torr) inside the hood where the whole process takes place. The substrate is positioned above the crucible containing the material, as is the thickness controller (quartz crystal scale). A shutter positioned between the metal crucible and the substrate is used to protect the substrate while the evaporation rate is regulated. Once the deposition rate has been controlled, the thickness controller is "zeroed", the shutter is opened and the PVD film manufacturing process begins until the desired thickness is reached. At this point the shutter is closed and the current is interrupted [24,26].

The PVD films of ML, EL, PL and BL lignins were manufactured using a Boc Edwards model 306 vacuum evaporator. Due to the limited amount of material available, it was decided to control each film deposition on the substrate using mass and thickness ratios. Approximately 5.0 mg were used in each deposition process, measured on an analytical balance, taking care to note the value of the thickness obtained in each deposition, which is read by a quartz crystal balance attached to the evaporator itself. Table I contains the masses weighed for the methanol, ethanol, propanol and butanol lignins. The intensity of the electric current used to heat the Ta crucible was around 1.95 A for all four lignins. This value was obtained from a thermocouple. It should be noted that different numbers of depositions were made due to the limited material available.

| Dep. | Lignin methanol | | Lignin propanol | | Lignin butanol | | Lignin ethanol | |
|---|---|---|---|---|---|---|---|---|
| | Pasta | Esp. | Pasta | Esp. | Pasta | Esp. | Pasta | Esp. |
| 1ª | 5,4 | 20,9 | 5,0 | 13,0 | 6,4 | 13,8 | 5,2 | - |
| 2ª | 5,6 | 17,9 | 5,5 | 15,8 | 5,6 | 12,8 | 5,4 | - |
| 3ª | 5,2 | 24,6 | 5,7 | 18,7 | 6,2 | 13,4 | 5,6 | - |
| 4ª | 5,4 | 17,7 | 5,5 | 21,1 | 5,6 | 12,5 | 5,3 | - |
| 5ª | 5.6 | 18,3 | 5,3 | 18,4 | 5,2 | 13,6 | 5,4 | - |
| 6ª | 5,2 | 21,2 | 5,5 | 18,4 | 5,4 | 14,7 | - | - |
| 7ª | 5,7 | 19,5 | 5,2 | 14,8 | 6,1 | 17,4 | - | - |
| 8ª | 5,4 | 17,7 | 5,3 | 13,8 | 5,1 | 13,5 | - | - |
| 9ª | 5,5 | 25,7 | 5,5 | 17,2 | - | - | - | - |
| 10ª | 5,6 | 26,1 | 5,3 | 18,1 | - | - | - | - |

Table 8.1: Masses in mg and thicknesses in nm for the evaporations of methanol, propanol, butanol and butanol lignins

## 8.2 UV-vis and FTIR

The growth of the PVD films of the four lignins was monitored by UV-vis absorption using a Varian Cary 50 spectrophotometer scanning from 190 to 800 nm and using a quartz slide with no film deposited as a reference (baseline). The study of film loss to water after successive washings of the film was also monitored by UV-vis spectroscopy. The washing process consisted of immersing the films in ultrapure water contained in a beaker, one by one, on a magnetic stirrer. Setting the stirrer in motion created a gentle vortex to which the substrate was subjected for 5 minutes. The FTIR measurements of the PVD films deposited on ZnSe substrates were carried out on a Bruker Vector 22 spectrometer, with a spectral resolution of 4 $cm^{-1}$ and 128 scans in transmission mode. Optical microscopy measurements were carried out using a Leica microscope with a 50X objective.

# Chapter 9

# Results and discussion

## 9.1 UV-vis absorption spectroscopy

UV-vis absorption spectroscopy was used to verify the growth of the PVD films. Figures 1, 2, 3 and 4 show the UV-vis absorption spectra obtained for consecutive depositions of each lignin, the masses of which are given in Table I.

The growth of the PVD films was monitored by the maximum absorption band at around 280 nm, which is attributed to the $\pi \dashrightarrow \pi^*$ transition of the phenyl group [22, 25]. It is worth noting that the UV-vis spectra obtained for the PVD films are very similar to those obtained for Langmuir-Blodgett (LB) films [27], suggesting no thermal degradation of the lignins, at least in relation to the phenyl group. The insets in Figures 1, 2, 3 and 4 show that there is a linearity between the maximum absorbance value at 280 nm and the film thickness expressed in nm (or the amount of material deposited in mg). This linearity indicates that with each evaporation process a similar amount of material was deposited on the substrate.

Although the linearity observed for the absorbance as a function of the deposited mass or thickness of the PVD films (insets in Figures 1, 2, 3 and 4) indicates that when the absorbance of the PVD films is greater than the mass of the PVD films, the absorbance of the PVD films is greater than the thickness of the PVD films.

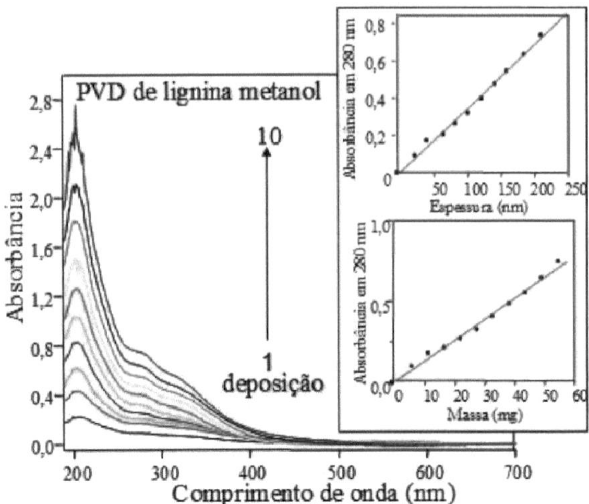


Figura 9.1: UV-vis absorption spectra of methanol lignin PVD film made from successive depositions on quartz sheets. Upper inset: absorbance at 280 nm vs thickness in mg of deposited material. Lower inset: absorbance at 280 nm vs mass in nm.

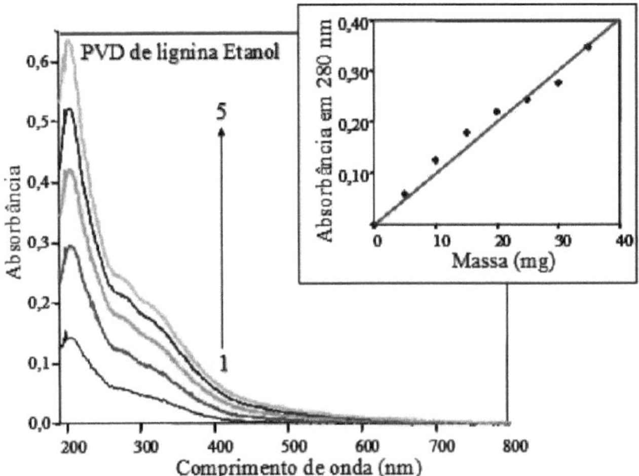


Figura 9.2: UV-vis absorption spectra of ethanol lignin PVD film made from successive depositions on quartz sheets. Inset absorbance at 280 nm vs mass in mg from successive depositions on quartz sheets.

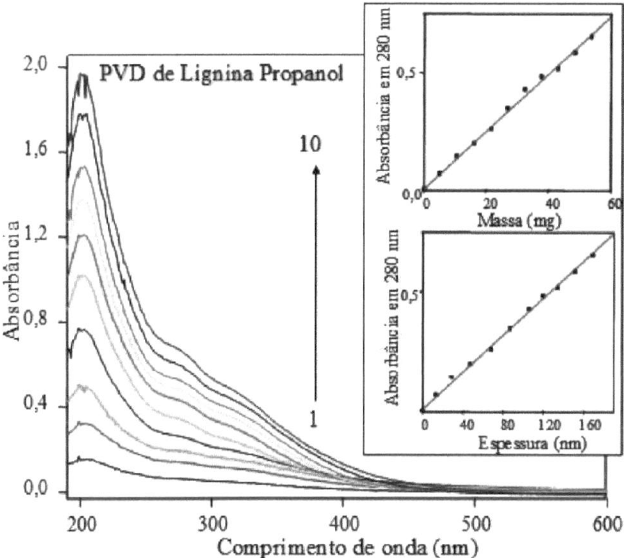


Figura 9.3: UV-vis absorption spectra of the propanol PVD film made from successive depositions on quartz sheets. Upper inset: absorbance at 280 nm vs mass in mg of deposited material. Lower inset: absorbance at 280 nm vs thickness in nm.

The fact that fairly close quantities of material are evaporated with each deposition process does not guarantee that the material is deposited homogeneously on the substrate. Therefore, optical microscopy measurements were carried out to make inferences about the morphology of these PVD films on a microscopic scale. Four images are shown in Figure 5 referring to the PVD films of methanol, ethanol, propanol and butanol lignins made with different numbers of depositions on quartz substrates. It can be seen that on a microscopic scale the films obtained are quite homogeneous, which is an important factor in terms of the reproducibility of the results when applying these films as sensors, for example.

With a view to the future application of these films in sensory units in liquid systems, it was checked whether there was any loss of material during the washing process, which is also an important factor in guaranteeing the reproducibility of the results when applying these films. The PVD films of the four lignins with the maximum number of depositions made on a quartz substrate were subsequently washed in ultrapure water following the procedure adopted when the four lignins were deposited on a quartz substrate.

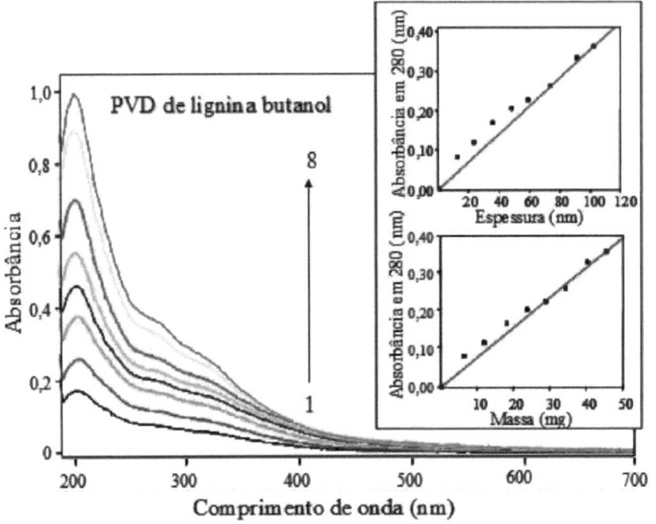


Figura 9.4: UV-vis absorption spectra of butanol lignin PVD film made from successive depositions on quartz sheets. Upper inset: absorbance at 280 nm vs thickness in mg of deposited material. Lower inset: absorbance at 280 nm vs mass in nm.

when working with sensors in liquid systems. Before and after each wash, a UV-vis absorption spectrum of the film was obtained, which is shown in Figure 6. It is possible to notice a significant loss of material during the wash, which, in principle, would make it impossible to apply these PVD films as sensing units if you wanted to reuse the sensors after a certain measurement. However, this loss most likely occurs because the film is relatively thick (many depositions -→ 150 to 250 nm), leading to a loss of material closer to the surface, i.e. further away from the substrate. There is a tendency for the loss of material to stabilize towards the "layers" closest to the substrate (lower absorbance). Considering that in the case of the sensory units the PVD films are deposited with a thickness of around 10 nm, the expectation is that the loss will be very small, as already observed for the LB films [27]. However, new tests will be carried out with 10 nm thick films to prove this possibility.

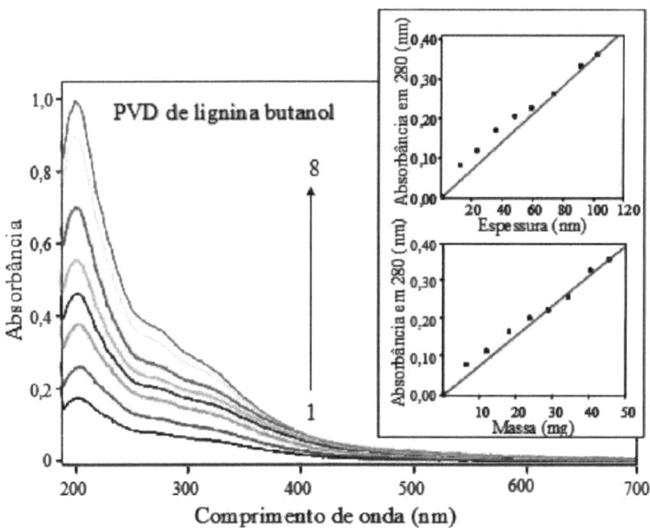


Figura 9.5: (a), (b), (c) and (d) optical microscopy images of PVD films of methanol (210 nm), ethanol (195 nm), propanol (170 nm) and butanol (112 nm) lignins deposited on a quartz slide.

## *9.2* Infrared absorption spectroscopy

FTIR absorption spectroscopy measurements were carried out in order to investigate the thermal stability of the lignin molecules subjected to the vacuum evaporation process to form the PVD films. Figures 6, 7, 8 and 9 show, respectively, the FTIR spectra of the powder and PVD films of the lignins methanol, ethanol, propanol and butanol. The assignments of the main absorption bands are given in Table II [12].

The FTIR spectra obtained for the PVD films of the four lignins shown in Figure 6 and the main IR absorption bands observed are the same as those reported in the literature for lignins [27], in general, and for these lignins extracted from sugar cane via the supercritical process [12]. This guarantees chemical integrity during the PVD film manufacturing process. Furthermore, with the exception of ethanol lignin, the other lignins presented spectra whose relative band intensities differed when comparing the FTIR spectra of the powder and the PVD films for each lignin separately. As the powder spectrum refers to a system whose

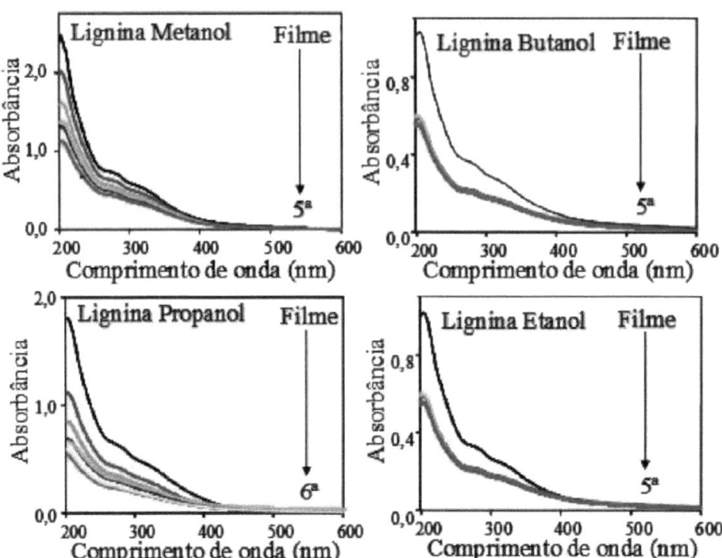


Figura 9.6: UV-vis absorption spectra, before and after successive washes, of PVD films of lignin methanol (210 nm), ethanol (195 nm), propanol (170 nm) and butanol (112 nm) deposited on quartz foil.

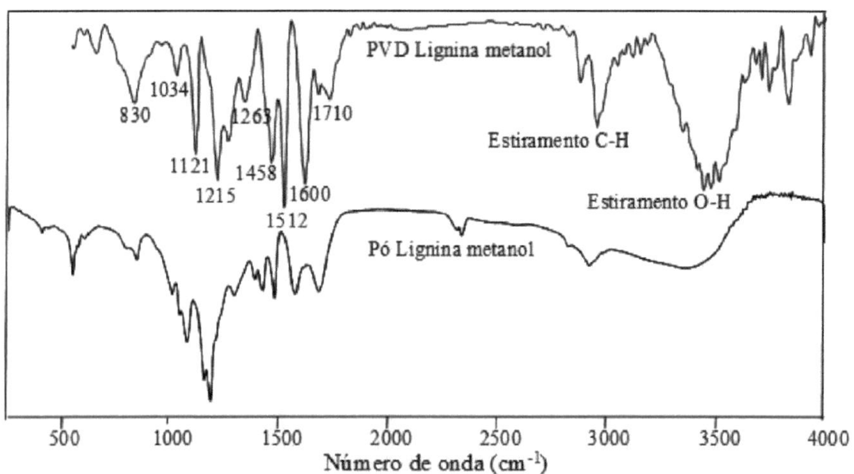


Figura 9.7: FTIR spectrum of methanol lignin PVD film (210 nm) deposited on ZnSe and FTIR spectrum of methanol lignin powder on KBr wafer.

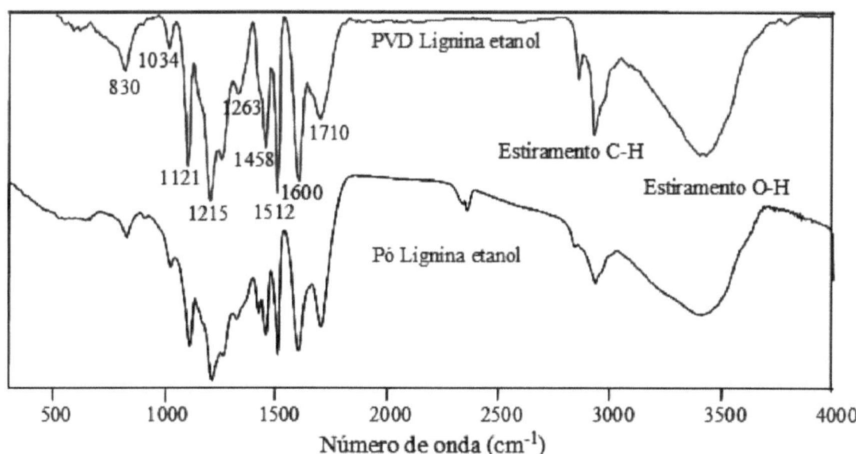


Figura 9.8: FTIR spectrum of ethanol lignin PVD film (195 nm) deposited on ZnSe and FTIR spectrum of ethanol lignin powder on KBr wafer.

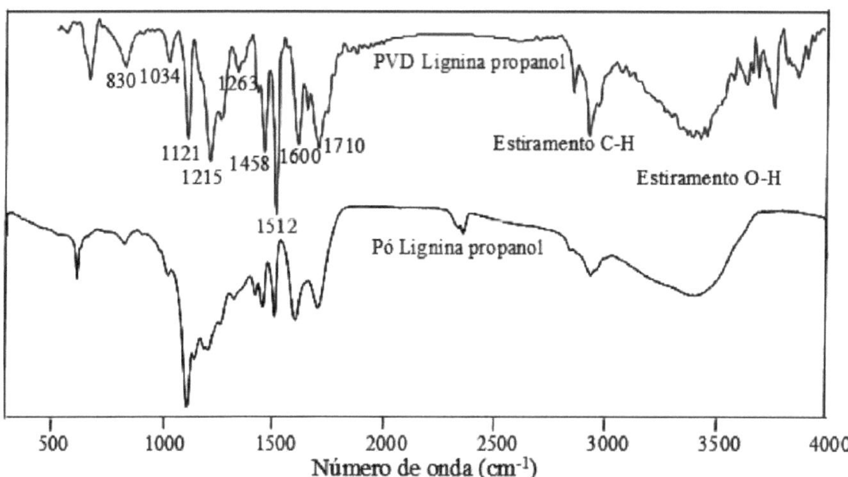


Figura 9.9: FTIR spectrum of ethanol lignin PVD film (195 nm) deposited on ZnSe and FTIR spectrum of ethanol lignin powder on KBr wafer.

molecules are randomly distributed spatially, these FTIR results suggest that methanol, propanol and butanol lignins are structured in an organized way in PVD films.

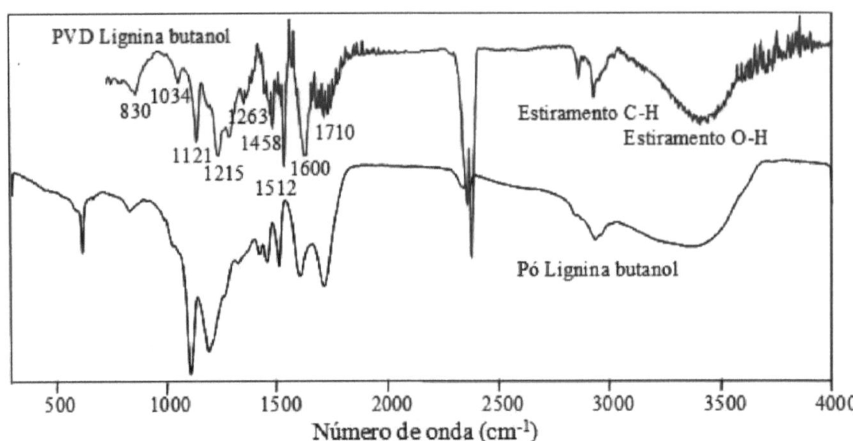


Figura 9.10: FTIR spectrum of butanol lignin PVD film (112 nm) deposited on ZnSe and FTIR spectrum of butanol lignin powder on KBr tablet.

| Center (cm 1) | Duties |
|---|---|
| 830 | Aromatic C-H (out-of-plane deformation) |
| 1034 | Aromatic C-H (in-plane deformation) and alcoholic C-O (deformation) |
| 1121 | C-O-C ether (deformation) |
| 1215 | C-O-C ether (deformation) |
| 1263 | Guaiacyl ring (deformation) |
| 1458 | C-H (deformation) |
| 1512 | Aromatic ring (C=C stretch) |
| 1600 | Aromatic ring (C=C stretch) |
| 1710 | C=O carbonyl (stretching) |

Table 9.1: Wave numbers and assignments for the main FTIR bands

The molecular organization can be determined by considering that the absorption intensity is given by the scalar product *I* = ~ *-~o,* where E is the electric field of the incident radiation and $\mu o$ is the variation produced in the dipole moment of the molecule by the incident radiation [12]. Finally, since for methanol, propanol and butanol lignins the FTIR spectra of PVD films are dominated by bands whose dipole moments oscillate in the plane of the molecule, generally associated with the C=C bond of the benzene ring [27], and since the electric field of IR radiation in the transmission mode is parallel to the substrate surface, it can be concluded that these lignins are preferentially positioned parallel to the substrate surface. In the case of ethanol lignin, the similarity between the FTIR spectra of the powder and the PVD film *suggests* a random structuring of this lignin in the PVD film. Figure 11a illustrates the selection rules, including the transmission and reflection modes, and Figure 11b schematically represents the preferential orientation of the methanol, propanol and butanol lignins, as well as the random orientation of the ethanol lignin. Comparing the molecular organization in PVD films of methanol, ethanol, propanol and butanol lignins with LB films of butanol lignin [12], it can be seen that this lignin is also preferentially oriented parallel to the substrate surface in LB films. In addition, we can find in the literature that LB films of lignins extracted from sugarcane bagasse with ethanol solvent and by saccharification (SAC) [11] exhibit an anisotropic molecular organization. The orientation of these lignins is perpendicular to the substrate for ethanol lignin and parallel to the substrate for lignin (SAC). It can therefore be seen that the molecular organization of the lignins depends not only on the technique used to prepare the thin films, but also on the process used to obtain these lignins.

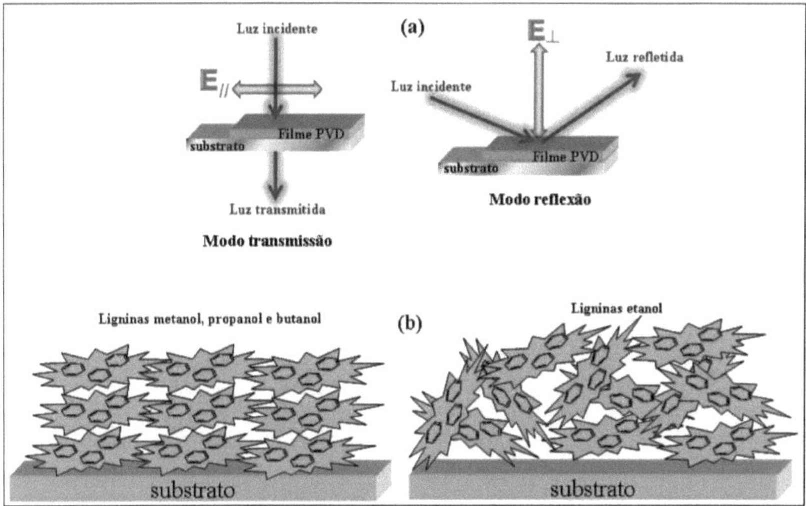


Figura 9.11: a) electric field and transmitted and reflected radiation in trans-mission and reflection modes, respectively; b) molecular organization of the lignins methanol,

ethanol, propanol and butanol.

# Chapter 10

# Conclusions

PVD films with controlled thickness on a nanometric scale were obtained using lignins extracted with the solvents: methanol, ethanol, propanol and butanol. The growth of the film with different thicknesses was monitored via UV-vis measurements. In morphological terms, the film was homogeneous on a microscopic scale. Using FTIR absorption spectroscopy, it was possible to investigate the molecular integrity of the lignin in the PVD film and verify that it does not undergo degradation with the temperature used to manufacture it. In terms of molecular organization, ethanol lignin is randomly structured in PVD films, while other lignins are preferentially structured parallel to the substrate surface in these films. The possibility of using PVD films of these lignins as sensory units for heavy metals in water and/or organic solvent vapors is currently being studied.

**Part II references**

1. Viscoelasticity of Biomaterials, Chapter 25, Internal Motions of Lignin, A Molecular Dynamics Study, T. Elder, School of Forestry, Alabama Agricultural Experiment Station, Auburn University, AL 36849.

2. P. Luner, U. Kempf, Tappi 53 (1970) 2069.

3. O.N. Oliveira Jr., C.J.L. Constantino, D.T. Balogh, A.A.S. Curvelo, Cellulose Chemistry and Technology 28 (1994) 541.

4. A.M. Barros, A. Dhanabalan, C.J.L. Constantino, D.T. Balogh, C.P. Neto, O.N. Oliveira Jr., Thin Solid Film 354 (1999) 215.

5. D. Pasquini, D.T. Balogh,O.N. Oliveira Jr., A.A.S. Curvelo, Colloids and Surfaces A: Physicochem. Eng. Aspects 252 (2005) 193.

6. G. Gilardi, A.E.G. Cass, Langmuir 9 (1993) 1721.

7. G.F. Martins, A.A. Pereira, B.A. Stracçalano, P.A. Antunes, D. Pasquini, A.A.S. Curvelo, M. Ferreira, A. Riul Jr., C.J.L. Constantino, Sensors and Actuators B 129 (2008) 525.

8. C.J.L. Constantino, L.P. Juliani, V.R. Botaro, D.T. Balogh, M.R. Pereira, E.A. Ticianelli, A.A.S. Curvelo, O.N. Oliveira Jr., Thin Solid Films 284 (1996) 191.

9. C.J.L. Constantino, A. Dhanabalan, A.A.S. Curvelo, O.N. Oliveira Jr., Thin Solid Films 329 (1998) 47.

10. C.J.L. Constantino, A. Dhanabalan, M.A. Cotta, M.A. Pereira-Da-Silva, A.A.S. Curvelo, O.N. Oliveira Jr., Holzforschung 54 (2000) 55.

11. D. Pasquini, D.T. Balogh, C.J.L. Constantino, P.A. Antunes, A.A.S Curvelo, R.F. Aroca, O.N. Oliveira Jr, Langmuir 18 (2002) 6593.

12. A.A. Pereira, G.F. Martins, P.A. Antunes, R. Conrrado, D. Pasquini, A.E. Job, A.A.S. Curvelo, M. Ferreira, A. Riul Jr., C.J.L. Constantino, Langmuir 23 (2007) 6652.

13. L.G. Paterno, C.J.L. Constantino, O.N. Oliveira Jr., L.H.C. Mattoso, Colloidal and Surfaces B: Biointerfaces 23 (2002) 257.

14. M. Norgren, S.M. Notley, A. Majtnerova, G. Gellerstedt, Langmuir 22 (2006) 1209.

15. S.M. Notley, M. Norgren, Biomacromolecules 9 (2008) 2081.

16. D. Hasegawa, Y. Teramoto, Y. Nishio, Journal of Wood Science 54 (2008) 143.

17. D.T. Balogh, A.A.S. Curvelo and R.A.M.C. de Groote, Holzforschung 46 (1992) 343.

18. MATTOX, D.M. Handbook of physical vapor deposition (PVD) processing. Noyes Publications, 1998.

19. TATSCH, P.T, 1996. V microelectronics workshop. Available at: < HTTP://www.ccs.unicamp.br/cursos/free107/dowload/cap11.pdf . Accessed on: April 13, 2008.

20. R.M. Silverstein, G.C. Bassler, T.C. Morril, Spectrometric identification of organic compounds, Guanabara Koogan (1994).

21. D. Campbell, J.R. White, Polymer characterization - physical techniques, Chapman & Hall (1989).

22. Skoog D.A.; Holler, F.J.; Nieman, T. A. Principles of Instrumental Analysis; Toronto, Bookman (2002).

23. J.B. Lambert, H.F. Shurvell, D.A. Lightner, R.G. Cooks, Organic structural spectroscopy, Prentice-Hall (1998).

24. P.A. Antunes, C.J.L. Constantino, J. Duff, R. Aroca, Applied Spectroscopy, 55 (2001) 1341

25. A.M. Bradshaw, E. Schweizer in Spectroscopy of Surface, R.J.H. Clark, R.E. Hester (Eds.), John Wiley & Sons (1988).

26. L. Gaffo, C.J.L. Constantino, W.C. Moreira, R.F. Aroca, O.N. Oliveira Jr., Journal of Raman Spectroscopy, 33 (2002) 833.

27. S.Y. Lin, C.W. Dence, Methods in Lignin Chemistry, Springer-Verlag, Berlin (1992).

## Part III

# PVD Films of Methanol, Ethanol, Propanol and Butanol Lignin with Impedance Spectroscopy and Aniline Vapor Exposure Measurements

# Chapter 11

# Experimental procedure

In this scientific initiation project, ultra-thin films (nanometers thick) of lignins extracted from sugar cane bagasse will be manufactured using the physical vapor deposition (PVD) technique, with the aim of applying these films as sensory units. The main information obtained in this project refers to the possibility of manufacturing sensors for heavy metal ions in aqueous solution. Film growth is monitored by ultraviolet and visible absorption spectroscopy (UV-vis). In the last report [1], PVD films were made from four lignins extracted with different alcohols (methanol, ethanol, propanol and butanol) and the results showed that films of these lignins do not undergo thermal degradation. Therefore, in this work, these four lignins were deposited on interdigitated circuits forming sensory units. These sensory units were characterized by impedance spectroscopy by immersing them in ultrapure water. However, the lignins did not resist the washing process, so a continuous loss of film was recorded. From this point on, with the non-reproducibility of three of the four lignins, the work converged on the manufacture of optical solvent vapor sensors [2], where the PVD film of propanol lignin was exposed to aniline vapor in order to detect whether the latter could be observed through UV-vis measurements, composing a new sensor. This IC project was carried out at the DFQB of the FCT/UNESP in Presidente Prudente and had the collaboration of professors D. Pasquini of the Federal University of Uberlândia A. Aprigio S. Curvelvelo. Aprigio S. Curvelo from IQSC/USP, who were responsible for extracting the lignins.

## 11.1 Introduction and justification

This report describes the activities carried out between September 2009 and May 2010 working with lignins extracted from sugarcane bagasse using the solvents methanol, ethanol, propanol and butanol. The following activities were planned for this period: i) prepare sensory units formed by depositing PVD films of lignins on interdigitated electrodes for the four lignins; ii) check the adhesion of the PVD films on the interdigitated electrodes via impedance spectroscopy by immersing the sensory units in water, as is done when applying the sensory units; iii) applying the sensory units to detect $Cu+^{2}$ , $Cd+^{2}$ and $Pb+^{2}$ ions in water at different concentrations via impedance spectroscopy; iv) depositing PVD films on ZnSe or Ge to investigate the lignin/metallic ion interaction via FTIR; v) investigating the effect of the lignin/metallic ion interaction on the morphology of PVD lignin films on a nanometric scale via AFM and on a micrometric scale via optical microscopy. Objectives (i) and (ii) were met and it proved not to be feasible to continue the work, as the interdigitated electrodes lost the lignin PVD film coating in the water washing process. However, the literature [2] shows PVD films of lignin extracted with ethanol alcohol forming optical sensors for solvent vapours such as aniline, so the work proceeded with new objectives i) manufacture of PVD films of lignin extracted with propanol alcohol; ii) exposure of these films to aniline gas vapour at different times; iii) characterization of these sensors via UV-vis spectroscopy; iv) characterization via Raman spectroscopy. These new objectives were fully met for one lignin. Section II provides theoretical concepts on PVD films, UV-vis absorption spectroscopy, Raman spectroscopy and impedance spectroscopy. Sections III and IV describe the experimental procedure and the results and discussion, respectively. Section V contains the conclusions of the work on the sensory units. The interest in studying lignins and the importance of manufacturing thin films has already been discussed in other reports [1,3].

# Chapter 12

# Theoretical concepts

## 12.1 PVD films

The physical vapor deposition (PVD) process refers to a set of deposition methods in which the material is deposited atom by atom or molecule by molecule. The materials are vaporized from a solid or liquid source and are transported in vapor form through a low-pressure atmosphere, solidifying on contact with the substrate [4]. PVD methods normally allow high deposition rates without causing damage to the substrate surface due to the low energy of the incident species. PVD processes usually have a thickness monitoring system that acts during the process (piezoelectric crystal) [5]. PVD processes are typically used to deposit films with thicknesses of a few nanometers to micrometers and can also be used to produce many layers. The PVD process can be carried out by vacuum evaporation, sputtering, voltaic arc deposition and others.

Vacuum evaporation, used in this work, is a type of PVD process in which the material vaporizes from a thermal source and reaches the substrate with little or no collision with gas molecules in the space between the source and the substrate. Vacuum evaporation can be carried out by resistive heating in which the source material is placed in a metal crucible or in a tungsten filament. The support is then heated by the Joule effect, melting the source material. Although very simple, evaporation by resistive heating has some restrictions depending on the material to be evaporated: refractory metals and many polymers cannot be evaporated due to their high melting point; evaporation of the filament material can contaminate the film and it is not possible to precisely control its thickness or alloy composition [5].

## 12.2 UV-vis absorption spectroscopy

UV-vis absorption spectroscopy is a technique that allows us to characterize materials by the incidence of electromagnetic radiation at different frequencies, which can lead molecules to an excited state of energy by absorbing the radiation. Electronic transitions involve the absorption of electromagnetic radiation whose energy is in the visible or ultraviolet region, UV-vis spectroscopy. The frequency of the absorbed radiation is a measure of the energy required for the electronic transition, while the intensity depends on the probability of this transition occurring [6-8]. In film research, the UV-vis technique is widely used to verify the growth of films, since, for a deposition with controlled thickness, the absorbance must increase linearly with the increase in the number of layers deposited, following Beer's law [9]. Determining the group responsible for absorbing UV-vis radiation and the wavelength range in which absorption occurs are also important pieces of information [9].

UV-vis spectroscopy measurements can also be used to observe the adsorption of solvent vapors to films, forming sensory units. This procedure was used in this work with the aniline solvent vapor. This solvent was chosen because it has harmful effects on health, especially when inhaled, and because it has been reported that vegetable extract films have good sensitivity to molecules with activated benzene rings [10].

## 12.3 Impedance Spectroscopy

Impedance spectroscopy measurements are a powerful tool for investigating interactions between analytes and the materials that make up the sensory units [11, 12,13]. The impedance spectroscopy technique basically consists of making measurements using an ac alternating voltage at the input of the circuit under study (sensory units in our case) in order to obtain capacitance and conductance values as a response [13]. Impedance scanning takes place over a relatively wide frequency range (generally between 1 Hz and 1 MHz), in which conductance can be easily converted into dielectric loss (G/w), allowing the detection of dispersions characteristic of the materials covering the interdigitated electrodes (LB or PVD films in our case), as a result of variations caused by the substances adsorbed on these materials [14]. Taylor and MacDonald [15] present an equivalent electrical circuit that describes the electrical characteristics of a system very similar to the one we are studying, i.e. a metal electrode covered with a weakly conductive material immersed in an electrolyte, represented schematically in Figure 2.

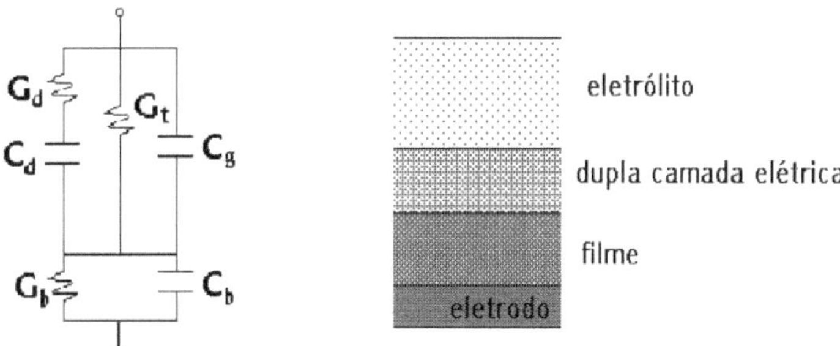


Figura 12.1: equivalent circuit of a metal electrode covered with a weakly conductive material, immersed in an electrolyte.

In this equivalent circuit, the presence of the film covering the electrodes is achieved by combining $Cb$ and $Gb$ *in* parallel. These two components are in series with the impedance of the electrolyte, which is made up of three components: the geometric capacitance of the arrangement of electrodes used immersed in an electrolyte ($Cg$); the electrical double layer ($Cd$) which arises due to the spontaneous adsorption of ions on the electrode/electrolyte interface, which is charged through the conductance of the solution ($Gd$); and the total conductance of the electrolyte represented by $Gd + Gt$. $Gt$ in this case allows charge transfer across the film/electrolyte interface [15]. Figure 3 is a simulation of the global dependence of capacitance (C) and dielectric loss ($G/\omega$) over a wide frequency range of the equivalent circuit shown in Figure 2.

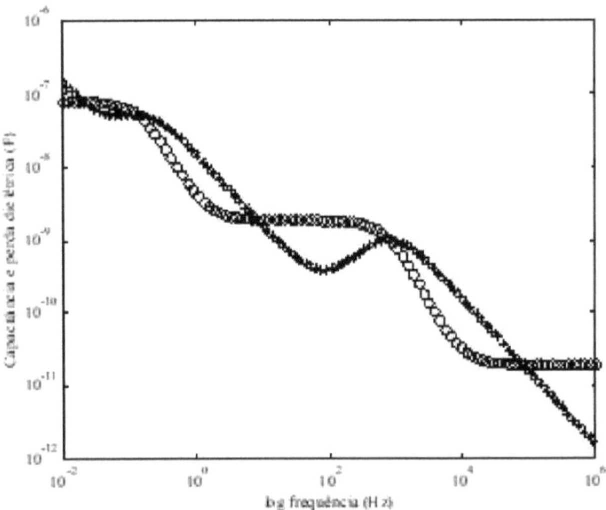


Figura 12.2: theoretical curves of C ($^{o}$ ) and *G/ω* (*) for the equivalent circuit diagrammed in Figure 12.1

In the curves illustrated in Figure 2, the low frequency region (< 50 Hz) is dominated by the effects of the electrical double layer, while the effects of the film covering the electrodes appear in the region between 102 Hz and 104 Hz. At frequencies above 105 Hz, the system's impedance is dominated by the geometric capacitance of the electrode configuration adopted. Ac measurements also avoid the unilateral displacement of ionized chemical species within the film, which could irreversibly alter its electrical properties, and are therefore a repeatable and non-destructive analysis process. Another advantage offered by ac measurements is the short time needed to vary experimental parameters [16].

# Chapter 13

# Experimental procedure

The lignin extraction process used in this work was described in the research project. Here we summarize the data on the lignins studied, which were extracted using methanol, ethanol, propanol and butanol with the following molar masses in g/mol: methanol ($Mn$ = 1150, $Mw$ = 1818, $Mw/Mn$ = 1.58), ethanol ($Mn$ = 1185, $Mw$ = 1971, $Mw/Mn$ = 1. 66), propanol ($Mn$ = 1082, $Mw$ = 1647, $Mw/Mn$ = 1. 52) and butanol ($Mn$ = 1062, $Mw$ = 1594, $Mw/Mn$ = 1. 50).

## 13.1 PVD film production

PVD films are made by evaporating the powder in a metal, cylindrical crucible, which has a small hole at the top through which the evaporated material escapes. The crucible is heated by the passage of an electric current, the intensity of which is generally no more than 2.0 A (10 V), which implies temperatures of less than 200° C for Ta (Tantalum) crucibles. The evaporation process is facilitated by the vacuum (≈ 10- 6- Torr) inside the hood where the whole process takes place. The substrate is positioned above the crucible containing the material, as is the thickness controller (quartz crystal scale). A shutter positioned between the metal crucible and the substrate is used to protect the substrate while the evaporation rate is regulated. Once the deposition rate has been controlled, the thickness controller is "zeroed", the shutter is opened and the PVD film manufacturing process begins until the desired thickness is reached. At this point the shutter is closed and the current is interrupted [17,18].

The PVD films of methanol, ethanol, propanol and butanol lignins were manufactured using a Boc Edwards model 306 vacuum evaporator. Approximately 5.0 mg were used in each deposition process, measured on an analytical balance, taking care to note the value of the thickness obtained in each deposition, which is read by a quartz crystal balance attached to the evaporator itself. The films were evaporated until they reached a thickness of 10nm. The intensity of the electric current used to heat the Ta crucible was around 1.95 A for all four lignins. This value was obtained from a thermocouple.

## 13.2 UV-vis

The growth of the PVD films of the four lignins was monitored by UV-vis absorption using a Varian Cary 50 spectrophotometer scanning from 190 to 800 nm and using a quartz slide with no film deposited as a reference (baseline). The study of film loss to water after successive washes of the film was also monitored by UV-vis spectroscopy. The washing process consisted of soaking the films in ultrapure water contained in a beaker, one by one separately, and on a magnetic stirrer. Setting the stirrer in motion created a gentle vortex to which the substrate was subjected for 5 minutes. These films were exposed to aniline vapor (1.0 mg/mL solution in water) at different time intervals, 20s, 120s and 20min, with a view to their application in the optical sensing of aniline vapors.

## 13.3 Impedance spectroscopy measurements

The electrical measurements (capacitance and resistance) were carried out using a Solartron 1260A impedance analyzer. Initially, measurements were made to analyze the geometry of the electrodes, where sweeps of capacitance versus frequency of alternating voltage were made with five interdigitated platinum electrodes immersed in water, which proved to be reproducible. Four electrodes were then used as substrates for PVD films of the lignins extracted with different alcohols, generating four films. One electrode was kept without a film (reference) to observe the changes caused by the film in the electrical response of the interdigitated electrodes.

Electrical measurements were then carried out in ultrapure water. The electrodes were first immersed in a 50 mL beaker filled with ultrapure water. We waited 20 minutes for the electrical signals from the electrodes to stabilize and then began the impedance measurements. We carried out thirteen sweeps for each sensor in ultrapure water with the alternating voltage varying in a frequency range between 1 Hz and 1 MHz.

# Chapter 14

# Results and discussion

## 14.1 Impedance spectroscopy

Initially, a study was carried out involving several electrical measurements for some Pt electrodes. This work was carried out to ensure that the geometry of the interdigitated electrode did not influence the measurements due to the high sensitivity of these electrodes. Subsequently, 10 nm of each lignin was deposited on the interdigitated electrodes. The results are shown below for each lignin, where a comparison is made of the electrode without film, and the respective measurements in water of the electrode with film at the maximum time of exposure to water. It is important to mention that for each measurement in water, the capacitance values were recorded by frequency, however, as the volume of results was very large, we chose to highlight the final standardized result as described above. The number of measurements in water varies with the stability of the measurements, i.e. the more film is desorbed from the electrode, the more measurements are made until the film no longer leaves the electrode.

Figure 3 shows the measurements of the PVD film electrode of methanol lignin, which was the lignin with the fewest measurements in water, four, as it did not show desorption.

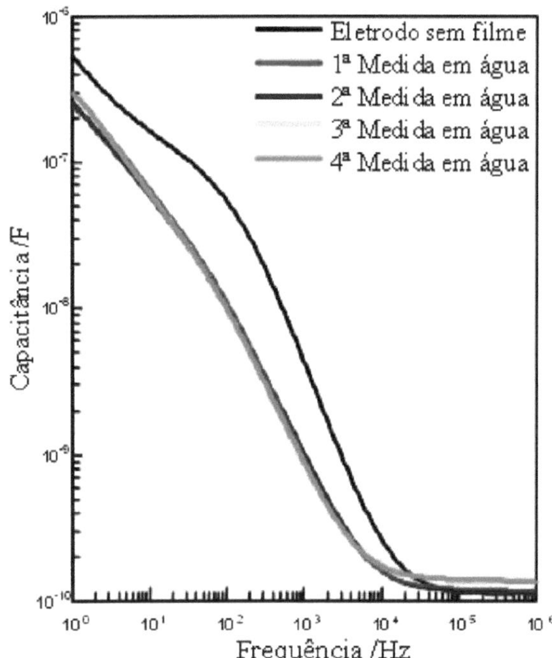


Figura 14.1: Capacitance measurements versus the frequency of alternating voltage for 2 interdigitated Pt electrodes in water for 20 min, one without a film deposited on the digits and the other with 10 nm of Iignins extracted with methanol.

film in water. This was the electrode with the best result. In general, the first measurements in water remove the excess film on the substrate created in the evaporation process, which delays the stability of the measurements. The remaining measurements confirm the adhesion of the remaining film to the substrate if the stability of the measurements is achieved. In the following measurements, Figures 4, 5 and 6, impedance spectroscopy revealed that the ethanol, propanol and butanol lignin films did not adsorb well when immersed in water. These results can be seen when looking at the curves in Figures 4, 5 and 6, which shift and do not reach stability.

The procedure for carrying out the measurements was modified to remedy this difficulty in the project. The magnetic stirrer was excluded from the measurements so that the vortex would not exert a force on the electrode, reducing the adsorption of the films. The

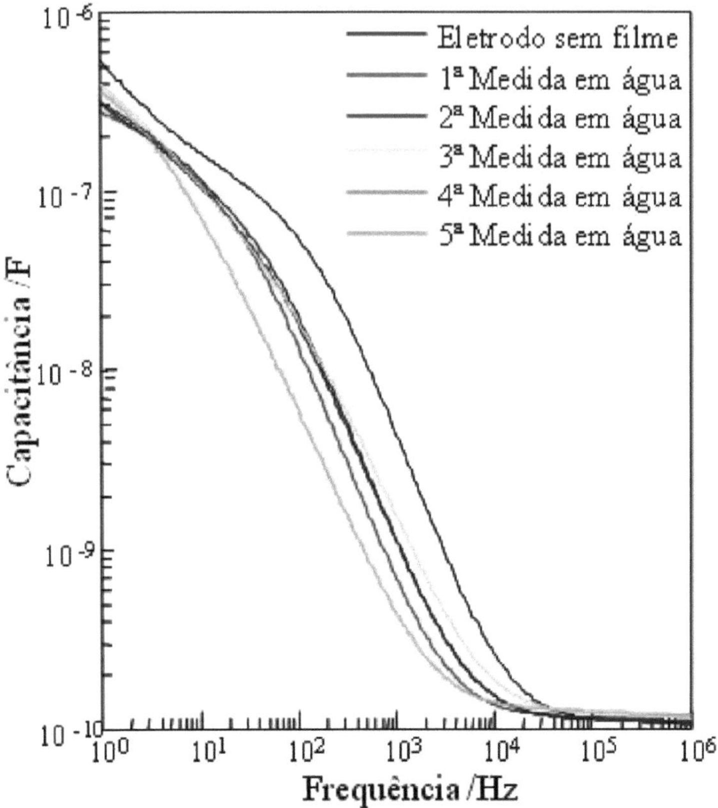


Figura 14.2: Capacitance measurements versus the frequency of alternating voltage for 2 interdigitated *Pt* electrodes in water for 20 min, one without a film deposited on the digits and the other with 10 nm of Iignins extracted with ethanol.

In other words, the magnetic stirrer was not the determining factor for the non-reproducibility of the measurements. This stage led to a change in the work: impedance spectroscopy proved not to be viable for PVD films of lignins extracted from sugar cane bagasse, unlike LB films of butanol lignin, which were successful as a sensing unit for heavy metal ions [19].

## *14.2* UV-vis absorption spectroscopy and exposure to aniline vapor

A PVD film of lignin extracted with propane alcohol was made and characterized using UV-vis absorption spectroscopy. Next, as suggested in the literature [2], we exposed the film to aniline gas for different times, 20 seconds and 120 seconds, and analyzed it at a wavelength of 206 nm, as shown below:

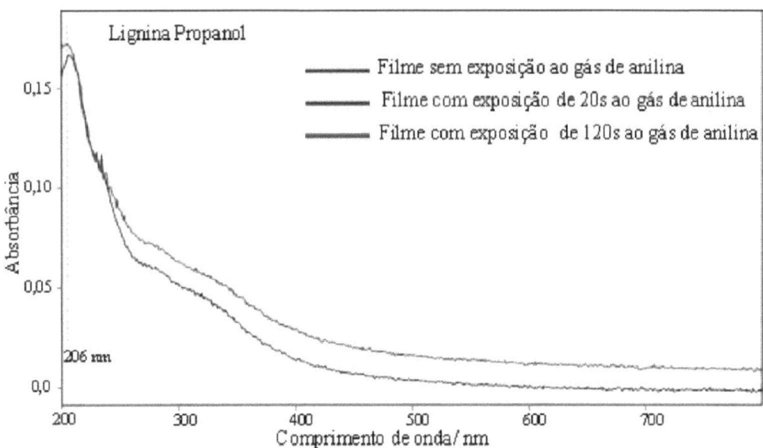


Figura 14.3: UV-vis absorption spectroscopy measurements for the propane lignin PVD film. Spectrum of the film at 20 s and 120 s of exposure to aniline gas.

Figura 14.4: Figure 7 shows that for times of 20 s the film did not absorb the aniline gas as the spectrum of the film did not change (superimposed curves). For times of 120 s there was a small variation at a wavelength of 206 nm. This shows that the film adsorbed the aniline gas. This is because vegetable extract films have good sensitivity to molecules with activated benzene rings [10, 11] and, since aniline is a good example of this type of molecule, the lignin absorbs the aniline causing an increase in absorption, with the maximum absorption band at around 280 nm being attributed to the $\pi \dashrightarrow \pi^*$ electronic transition of the phenyl group [8, 20]. Therefore, propane lignin can be used as an optical sensor unit for aniline vapors. The optical microscopy shown in Figure 8 shows the effect of gas adsorption on the morphology of propane lignin PVD films on a *micrometric* scale.

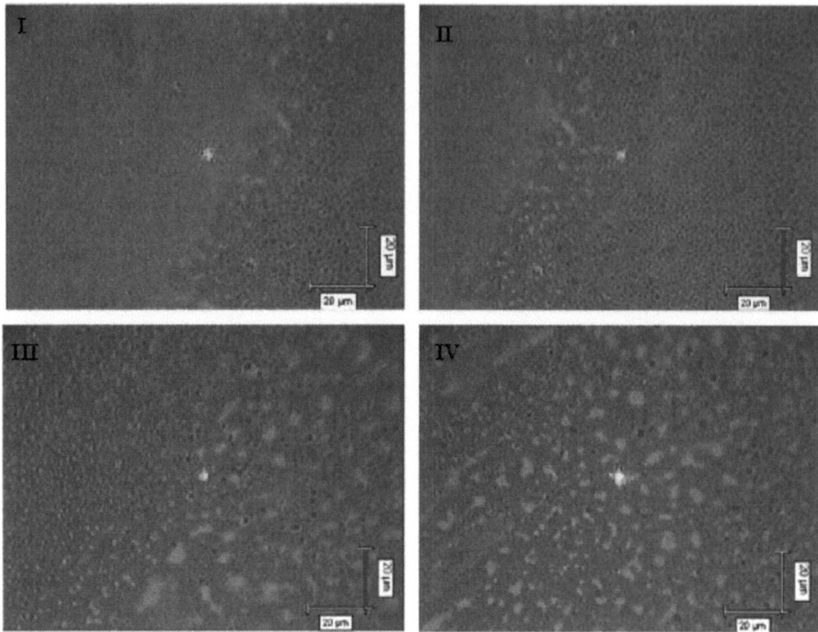


Figura 14.5: Optical microscopy measurements of different regions of the PVD film for a 20 min exposure time to aniline gas. Há image is close to the center of where the gas was most exposed in the sequence from I to IV.

Figure 8 shows that the closer to the central point where the greatest exposure to aniline gas occurred, there is an increase in the radius of the structures that have appeared on the film and a decrease in the density of structures, and the further away from this point there is an increase in the density of structures and a decrease in their radius, reaching a point where there are no more structures, where the film is homogeneous and has possibly not received any gas. We can understand this fact by thinking that the closer to the center where the gas hit the film, the greater the probability of the aniline molecules colliding with the same point, forming irregular structures with a large radius; the further away from this region, the probability of collision at the same point decreases, forming regular structures similar to circles, but decreasing the radius.

# Chapter 15

# Conclusions

PVD films using Iignins extracted with the solvents methanol, ethanol, propanol and butanol were deposited on interdigitated electrodes to evaluate the possibility of applying these films as sensory units. This expectation could not be realized because during the process of immersing the electrodes in water, three of the four films had poor adhesion to the substrate, as revealed by impedance spectroscopy, and consequently the film came off the electrode. The procedure was modified, but the result remained the same, showing that these films cannot be used as sensors for heavy metal ions in aqueous solution. As lignin is rich in aromatic rings, lignin extracted as propanol alcohol was tested as an optical sensor for solvent vapors, in this case aniline. This lignin proved to be a good sensor for times of 120s when analyzed by UV-vis absorption spectroscopy.

**References for Part III**


1. D.C. Ferreira and C.J.L. Constantino, Structural characterization of ultra-thin films of Iignins extracted from sugar cane manufactured by evaporation, Scientific Initiation Report, DFQB, 2009.

2. P. Aléssio, A.S. Cavalheri, D. Pasquini, A.A.S. Curvelo, C.J.L. Constantino, Revista Brasileira de Aplicaçoes de Vàcuo 27, (2008) 169.

3. D.C. Ferreira and C.J.L. Constantino, Structural characterization of ultra-thin films of lignins extracted from sugar cane bagasse manufactured by evaporation, Scientific Initiation Report, DFQB, 2008.

4. MATTOX, D.M. Handbook of physical vapor deposition (pvd) processing. Noyes Publications, 1998.

5. TATSCH, P.T, 1996. V microelectronics workshop. Available at:

< HTTP://www.ccs.unicamp.br/cursos/free107/dowload/cap11.pdf . Accessed on:

April 13, 2008.

6. R.M. Silverstein, G.C. Bassler, T.C. Morril, Spectrometric identification of organic compounds, Guanabara Koogan (1994).

7. D. Campbell, J.R. White, Polymer characterization - physical techniques, Chapman & Hall (1989).

8. Skoog D.A.; Holler, FJ.; Nieman, T. A. Principios de análisis instrumental; Toronto, Bookman (2002).

9. J.B. Lambert, H.F. Shurvell, D.A. Lightner, R.G. Cooks, Organic structural spectroscopy, Prentice-Hall (1998). 10. Y.E. Silina, T.A. Kuchmenko, Y.A.I. Ko-renman, O.M. Tsivileva, V.E. Nikitina. Journal of Analytical Chemistry 60 (2005) 678.

11. D. Savitri, C. K. Mitra, Bioelectronics and Bioenergetics 48 (1999) 163.

12. K. Toko, Meas. Sci. Technol. 9 (1998) 1919.

13. J. R. Macdonald, Impedance spectroscopy; New York; John Wiley & Sons; 1987.

14. A. Riul Jr, A. M. Gallardo, S.V. Mello, S. Bone, D. M. Taylor, L.H.C. Mattoso, Synthetic Metals 132 (2003) 109.

15. D. M. Taylor, A. G. MacDonald, J. Phys. D. Appl. Phys. 20 (1987) 1277.

16. A. Riul Jr, D. S. dos Santos Jr, K. Wohnrath, R. di Tommazo, A. A. C. P. L. F. Carvalho, F. J. Fonseca, O. N. Oliveira Jr, D. M. Taylor, L. H. C. Mattoso, Langmuir 18 (2002) 239.

17. P.A. Antunes, C.J.L. Constantino, J. Duff, R. Aroca, Applied Spectroscopy, 55 (2001) 1341

18. L. Gaffo, C.J.L. Constantino, W.C. Moreira, R.F. Aroca, O.N. Oliveira Jr., Journal of Raman Spectroscopy, 33 (2002) 833.

19. S.Y. Lin, C.W. Dence, Methods in Lignin Chemistry, Springer-Verlag, Berlin (1992).

20. A.M. Bradshaw, E. Schweizer in Spectroscopy of Surface, R.J.H. Clark, R.E. Hester (Eds.), John Wiley & Sons (1988).

Printed by Books on Demand GmbH, Norderstedt / Germany